THE SECRET LIFE OF STREAMS

by Lynell Marie Garfield

THE SECRET LIFE OF STREAMS
by Lynell Marie Garfield

www.clearmountainstream.com

Published by Lucky Bat Books
www.luckybatbooks.com
Reno, Nevada USA

Copyright © 2013 by Lynell Marie Garfield
Story illustrations by Daniel Devine
Field guide illustrations by Jason Barnes
Final design and layout by Sandra L. Silva

All rights reserved. No part of this book may be reproduced or transmitted to any form or by any means, electronic or mechanical, including photocopying, recording, or by any information and retrieval system, without written permission by the Author.

Library of Congress Cataloging-in-Publication Data

Garfield, Lynell Marie
The Secret Life of Streams / by Lynell Marie Garfield; Illustrated by Daniel Devine.
(First Edition) – ISBN 1-9390-5133-9
ISBN 978-1-939-05133-2

Printed in The United States of America

Dedication

This story was born to reconnect us to the waters.

For Kassidy and Taylor,
my beautiful inspirations to write this tale.

For Thomas, my Twin Flame,
for believing in me and touching the spark to my paper.

For August, I thank you for the energy you gave me to craft this book.

May your love for the oceans, rivers, and creeks go on forever.

I Love You!

Cast of Characters:

Loralei - the swimming mayfly

Apollo & Diana - the clinging mayfly twins

Rocky - the stone case-building caddisfly

Kassidy - the free-swimming caddisfly

Nettie Taylor - the net-spinning caddisfly

Madge - the midge fly

Grand Daddy - the giant stonefly

Ms. Dobbie - the hellgrammite

Gus - the dragonfly

Small Fry - the trout kids

Big Bad Voodoo Trout - the trout kids' parents

**PSST. Hey, you!
Yes, You.**

Hi there! My name is Loralei the Mayfly.

I know, you think you're dreaming right now, but you're not. It's ok, don't be shy. I'm perfectly harmless. I just want to be friends, and I've got some great secrets to share with you!

Like what?
Well, for starters, what if I told you there are creatures living near your house that are so odd they seem like underwater aliens? And what if I told you I could introduce you to them? I can, because they are all my friends! There are lots of strange but friendly creatures right here in your neighborhood stream, and I'll bet you didn't even know we were here! We live beneath the surface of the water -
you could say we are undercover.

Any day now, I will be leaving my underwater home. My wing pads are getting bigger and bigger, and my Auntie Madge says soon they will turn into wings and I will hatch out of the water. While I'm still here, I want to introduce you to all of my neighborhood friends here in Small Creek. Then, when I grow wings and can fly, you could introduce me to your human friends in the above-water world.

Is it a deal? Ok, then come underwater with me!

I live right here, in Small Creek just behind your house. You've probably never seen a mayfly kid before, huh?

That's funny, because I see humans at the edge of our watery world all the time. This is the first time I've talked to one, though. I'm so tiny and you are so big! But I'll bet we could still have a lot of fun together. I have some great friends and I can't wait to share my world with you.

I have lived in this creek my whole life and am excited to leave my home to try something new. After I hatch into the air, I'll get to use my new wings and explore the above-water world. Still, I am a little nervous to leave my wonderful friends and family here in Small Creek. I am excited to go live in a strange new place, but it will be weird. Almost as weird as it would be for you to suddenly live beneath the water!

First, My mom laid many tiny eggs at the edge of Small Creek. One of those was me! All of us eggs fell into the water and sank to the bottom of the creek. Nestled in the rocks at the bottom, I grew inside my egg and soon hatched out into the water.

I crawled out of my egg as a tiny nymph or *larva*. I was a young mayfly with legs but no wings - a cute little wormy bug. I grew in Small Creek for nearly two years, getting bigger and shedding my skin as it got too small.

This is me, now, as a mayfly nymph!

When it gets to be time for me to hatch, my wing pads will be dark and a little bigger. When they break open, I will leave the bottom of the creek, where I feel safe, and rise up to the surface. There I'll sit on top of the water, unfurl my shiny new wings and dry them out, fanning them slowly. After they dry, I'll fly away, heading upstream to lay my own eggs at the water's edge.

The cycle of life will begin all over again!

This is my home, Small Creek. Like I said, I have lived here my whole life.

There are a lot of things about us water dwellers that might surprise you. I'll bet you didn't even know mayflies like to nibble on blackberries! Twice now, I've seen the leaves turn yellow and red, and I've seen berries hanging over the edge of the creek get fat and juicy and drop into the water (yum!).

Our worlds are connected in more ways than you might know. For instance, when you and your friend were building that rock bridge across the stream last summer, you may not have thought about it, but it cut my neighborhood in half! I couldn't swim down to see many of my friends until the rains filled up our creek and overflowed those rocks. You see, we are here even if we are too small for you to easily see. Like I said, we are all connected, even if we are very different.

We do affect each other with many of the choices we make. You, me, and we all share water, air, and sunlight. I hope we can be friends and help each other, because your world up there is so beautiful and so is ours.

I know you are probably curious about how different we are from you. We are different from each other, too! Even though we come in different shapes and sizes, most of my friends and I are all young aquatic insects called *larvae*. As stream bugs, we spend nearly our whole lives underwater, and we each live in a special place and do things in our own special way.

To fit into each of our own different places in the creek, or habitats, we have each had to figure out how to survive the easiest. That means we have had to adopt different habits that work best for us to fit in and make our living where others can't. Let me tell you, the competition is wicked! There are waters that are shady, deep, and cold with lots of trees covering the water (a) or sunny, shallow, and warm (b). There are rocky reaches with fast, gurgling water (a/b) or muddy, slow-moving waters that are thick with reeds (c).
There are stream bugs living in all of these very different spots!

No matter where we live, we all need to breathe, eat, stay on the stream bottom in moving water, and avoid getting eaten by the Big Bad Voodoo Trout.

Unlike you, we don't live in houses down here... the water of Small Creek is our home. We all live near each other in this neighborhood, no matter what we look like.

Even though we like very different places to live, we still watch out for each other and hang out together. We are just like a big, diverse, weird family. Big, because there are lots of us. Diverse, because we are all so different. Weird, because... well, you'll see!

If you listen to what the stream bugs have to tell you,
we will tell you the story of the stream.

Everything is not always as it seems. I am a kind of fly, a mayfly, but I've lived my entire life underwater, not flying anywhere! Auntie Madge told me once flies hatch and leave the water, we never look back. Most of us in this stream are like human teenagers,
waiting to get our "license to fly."

Whether you are a human child or a mayfly in Small Creek, breathing air is important to life. You may not think that we would need to breathe air because we live underwater, but we do! Some of us have things called gills to help us breathe oxygen in the water.
(See? Mine are along the edges of my belly)

It's important for us to have clean water to live in. Clean water carries oxygen to us underwater, but dirty water cannot. Some of us can handle living in mucky water and do well there. However, those of us who are sensitive stream bugs, absolutely need clean water.
If clean water becomes dirty, we sensitives may just leave town.

Mayflies like me, and most other nymphs here, live nearly our whole lives in the bottom of the stream before we hatch out into the air. I have cousins who burrow deep into the mud, some who climb on reeds, and even more who cling to rocks in fast-moving riffles where the water gurgles and rushes over rocks.

My favorite thing to do is go find some of the Small Fry, the trout kids, when the Big Bad Voodoo Trout are away. These kids are just a little bigger than me and live close to the gravel at the bottom of the creek, near the big deep pool.

There are lots of these guys; their parents sure must have been busy to have so many kids!

These little guys all look the same; they are silvery-colored with big bellies and one big eye on each side of their head. The older ones have a dark line on their sides and spots. We aren't really supposed to play together, since fish usually eat bugs. But I'm not afraid of them, and it's fun to swim with Small Fry!

You see, as a swimmer-mayfly, I just love to race! You can tell the way my body is shaped, very streamlined and smooth, that I am a fast swimmer. I flip my tail and move through the water just as quick as can be, even though I don't have fins like fish. The Small Fry are getting faster, but in our racecourse at the bottom end of the pool, I hold the record.

Who says girls can't beat boys?

As a swimmer, I see everybody else in the stream, which is great, because I get to visit with everyone! Some other bugs just hang out in their own little part of the neighborhood and don't know anyone besides the guy living right next door.

My twin cousins (Apollo and Diana) make their living by clinging to rocks, instead of swimming. The twins are called clinging mayflies, while I am a swimming mayfly. They like to hang out in the fast-moving water, clinging to the tops or bottoms of rocks. Their heads and bodies are very flat, perfect for blending in with the rocks so the water's flow goes right over them.

All of their six legs are really muscular, perfect for hanging onto rocks in the rapids.

I have a hard time swimming in the fast current, but Apollo and Diana just hang out with their faces in the bubbles, pressed flat against a big rock. Their thick, flat teeth are great for scraping algae off the rocks, and they say they love it. Yuck! That doesn't sound good to me... I'd much rather see what delicious things I can find to eat floating in the water.

My friend Rocky is a stone case-building caddisfly. The word "caddis" means case, and Rocky's case looks like a homemade sleeping bag made of pebbles.

The only way Rocky will ever climb out of his case is if he outgrows it! He has big hooks on his rear end that hold the case onto him tightly. This way, he drags it with him as he crawls around in the shallows, basking in the warm sun.

Rocky says his case protects him from the Big Bad Voodoo Trout, who love to eat stream bugs like us. Rocky can't outrun the big fish, so his sleeping bag is the next best thing to protect him. He just pulls his legs and head inside until trouble passes. After all, fish don't like to eat rocks!

Rocky's pebble-case is heavy enough that it helps him stay on the bottom of the creek where the algae grows. Like Apollo and Diana, he is really hooked on algae. He even has the same big teeth as the twins, for scraping algae off the rocks.

Rocky's favorite place is on the big rocks in the sun at the slow edges of the creek. This is where he finds the tastiest algae! Unlike Apollo and Diana, Rocky wouldn't be able to hang out in the fast water, because his body isn't flattened like theirs. Rocky moves slowly and always seems to be in about the same place. I would love to bring him on some of my adventures, but he's just too slow for me!

One of my favorite bugs to swim with is Rocky's cousin Kassidy. She's a caddisfly too, but she doesn't have a sleeping bag like Rocky. Instead, she is a free-swimming caddisfly, a little green worm with arms and no case. Kassidy and Rocky look so different, sometimes I wonder if they really are cousins.

Since Kassidy doesn't stick all those rocks to her body, she isn't weighted down on the bottom of the creek like Rocky is. She crawls around on the stream rocks at the bottom to find food. And when she loses her footing, she curls up in a little ball and tumbles downstream in the current.

Kassidy does a good job hiding from the Big Bad Voodoo Trout with her pretty green camouflage!
When she is below me in the leafy brown-green water, she's easy to see because she's light green.
But when I look up at her from the bottom of the creek, where the Trout live,
the glistening water and reflections of trees and clouds make her nearly invisible.
I love to watch her move from deeper water up to the surface; she makes
funny faces as she puffs up her belly with air and rises up to the top.

I tell her "Hey! Watch out for the Big Bad Voodoo Trout!" She usually just giggles,
winks at me from above, and tumbles in the water like a balloon
until she can grab ahold of another rock.

Another caddisfly cousin is Nettie Taylor, a net-spinning caddis who lives in the mud downstream from the dam. Nettie doesn't have a case, instead, she lives inside a net she has spun. The net catches food for her, while she just relaxes in the muddy water. There are lots of others like her in our stream, I think because we have lots of mud that spills through the dam. Nettie has figured out how to live well in the mud.

Nettie lives inside a crack in an old log downstream from our neighborhood and has spun her net right on the side of the log. She hangs out in that crack, and last I heard, she had spun another net just a little way away from her main one. She eats whatever goodies get stuck in the net. Her net looks strong enough that she could almost catch small fish. If only it were a bit bigger, we'd show those Trout, once and for all!

We all love to visit Nettie, to see what gets caught in that net. It's always something different!

Since Nettie lives in the muddy part of the creek and cannot move around like me, she needed to grow a different way to breathe underwater. I can swim around to find cleaner water if it gets too muddy for my taste. Nettie can't swim away, so she has hairy little gills along the long inside edges of her belly, which she waves around and vibrates while she's inside her net. Doing this actually creates a current and brings in more oxygen for her to breathe. This rhythm also draws floating food into her net, which is great for her!
I told her she should try belly dancing next. She'd be great!

Madge the midge fly was my mom's best friend before she hatched and left Small Creek. Madge has watched over me since I can remember. Madge is a hip old fly, but she gets nervous when I go down to play with the Small Fry.
I think she's afraid I'll get eaten…

Since my mom hatched out of the creek, Auntie Madge treats me like her own young fly, wagging her stubby proleg at me and warning me.
She says in her raspy voice,
"Loralei, get away from those Small Fry! You'll be sorry if their parents find you!"

I can sprint faster than the trout kids, but maybe not as fast as the Big Bad Voodoo Trout. I always laugh when she yells to me, but she may be right - they would see me as a good snack!

My Auntie Madge is very different from me, with only a proleg instead of my jointed legs. She has no wing pads, antennae nor tails. She always looks like she just woke up, with her antennae shrunk into her tiny head and her wormy body propped up from the muck. But then, all of her friends look that way. They hang out together playing cards and watching everything that happens. Madge is always bellowing from her muddy home. The only reason the big Trout don't have her for lunch is there are so many bigger bugs living down here.

Some bugs are much bigger than Madge and make great eating for the Trout, like the great stonefly Grand Daddy! Grand Daddy is a big guy, and spends his time hiding deep under the rocks. Grand Daddy says he's "meditating," but I am not sure what that means. I have heard rumors that he has friends who like young mayflies as snacks, so I'm not going to ask. He lives upstream, where the big trees hang over the water. He told me that I shouldn't be afraid of him, because he's a vegetarian, but then he does this weird thing with leaves that fall into in the water. When he eats the leaves, all that is left afterwards are their skeletons!

The other big bug around these parts is Ms. Dobbie. She lives upstream near Grand Daddy and has lots of feelers sticking out of the sides of her belly that look like legs. She is even bigger than he is, and very shy. I've only seen her twice. I think that's probably a good thing, because she has huge jaws and I'll bet she knows how to use them.

Ms. Dobbie hides from the big Trout by climbing under and burrowing beneath the rocks. If she needs to, she can still breathe under there by using her long gill-feelers like little straws sticking up into the clean water. I hear the human kids are afraid of her because she likes to bite people. That's probably how she got her nickname "toe-biter!"

Last of all, I want to tell you about Gus the dragonfly. We need to keep our eye on Gus, because he is not always the nicest bug. He's got a big belly, which is full of gills. He can breathe in the mud by moving the gills back and forth to bring air and water into his belly, kind of like Nettie, who also lives in mud. Sometimes I think he shows off to impress us.

Gus can shoot water out his rear end. Sometimes he looks like he's got jet propellers shooting him through the water. As you might guess, it can be a little gross to be the bug swimming behind him.

But it works! I've seen him escape from being eaten by one of the big Trout by blasting it with water and air. He confuses it and jumps forward to safety.

I am glad I can swim faster than Gus, because he likes to eat other flies. He buries himself in sediment, hiding and watching for food to come swimming by.

Last week, I saw Gus eat a huge diving beetle! Gus was mostly buried, hiding out in the mud just below the riffle, and the beetle swam too close to him. In an instant, he jumped out of the mud, grabbed the big beetle, and his mouth opened up so wide that he ate the beetle in three bites!

Now you have a glimpse into our secret world. Though I won't be able to go back and forth through the worlds once I sprout wings, you still can! I've shown you who to look for, and you can come out to the creek behind the houses and explore whenever you want. I know you've seen the bushes around the creek, even though it's pretty overgrown.

To find us, just come out to the creek and turn over rocks or stir up the gravel to visit the bug kids in your creek. Like you've seen, we hide in the water, so you may need to look hard, maybe even dig below the rocks to find us. By now, you probably understand why we hide from bullies. For us, the bullies are birds, housecats, and those pesky Trout. We try to stay out of sight of those hungry creatures.

Now that you've met my friends, once I hatch out, I can't wait to see your world and meet your friends! I'll need some new friends above the water. Until then, come out and play in Small Creek.

Loralei's Safety Tips to Follow at the Creek

It's always good to let your folks know before you visit the water.

Bring drinking water and a sun hat with you. Don't let the heat get the best of you!

Be sure to stay in shallow water and walk carefully on rocks; they might move or be slippery.

Keep your shoes on. Rocks in streams are sometimes smooth, but hidden glass or other sharp objects could give you a painful story to hobble home with.

Did you know?
Field guides are tools for exploring a creek or river by giving clues, or key body parts, to identify a *larva* or other critter.

FIELD GUIDE: Use these next pages to find me, Loralei, and my friends in YOUR creek!

Take your time! We are small (as small as this line -) to large (as large as the palm of your hand!)... we may be hard to find.

Carefully, walk into shallow water and begin to turn over rocks to see who is underneath them. When you pick up a rock out of the water, turn it over to inspect the bottom, and gently blow on it. We bugs and other creatures will wiggle to get out of the wind.

If your creek is muddy, you may need a small net for catching us when you stir up the mud. Remember to hold the net downstream of your feet, to catch what you stir up with feet and hands!

If you bring a magnifying glass and look closely, you might easily find some of my friends in your stream.

My Stream Field Guide

Creek Explorer Name: _____

Loralei
The Swimmer Mayfly

*Hook on feet, not pinching claws

*1 set of wing pads

Tails (2-3)

Round gills along sides of abdomen

*Unlike the stonefly

Apollo and Diana
The Cling-on Mayflies

6 strong legs

*1 set of wing pads

"Dorso-ventrally" flattened body (looks squished)

*Hook on feet, not pinching claws

*Unlike the stonefly

Grand Daddy
The Giant Stonefly

** Pinching claws on feet

Gills under 6 jointed legs
(looks like armpit hair)

** 2 sets of wingpads on back

2 tails

**Unlike the mayfly

Kassidy
The Case-less Caddisfly

Worm-type body
(no stone case like Rocky)

*6 jointed legs

White or green color

2 prolegs on posterior with hooks

*Unlike the midge fly

Rocky
The Stone case-building Caddisfly

*Slightly flattened head

*6 jointed legs sticking out of the front of the case

Case made of stones

*Usually visible unless he is scared and hiding inside the case

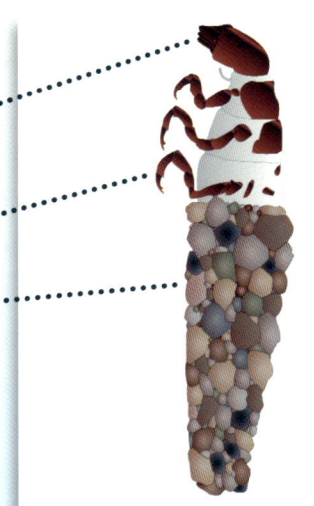

Nettie Taylor
The Net-spinning Caddisfly

**6 jointed legs

**Hairy gills along underside of belly and posterior end

**2 prolegs on posterior with hooks

**Visible if not hiding in net

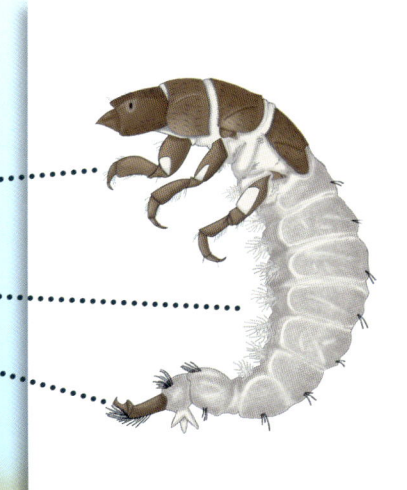

Gus
The Dragonfly

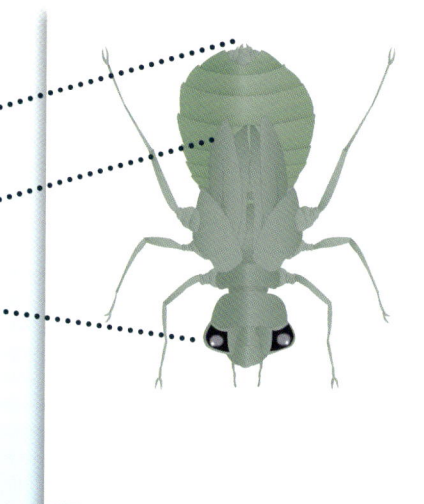

3 stiff, short, pointed "tails"

Stout, broad belly with visible segments

Large eyes on sides of the head

Mask-like lower jaw that telescopes open to eat larger critters

Dobbie
The Hellgrammite

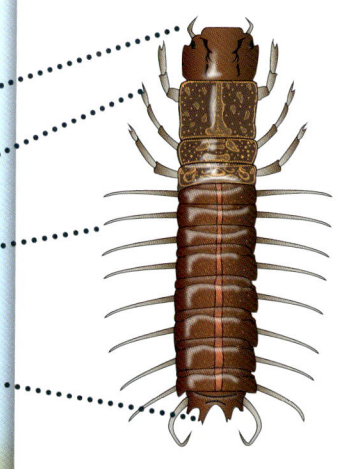

Big pinching jaws (up to 4 inches)

6 Jointed legs near head

8 pairs of lateral filaments along sides of belly

2 prolegs on posterior end with hooks

Madge
The Midge Fly

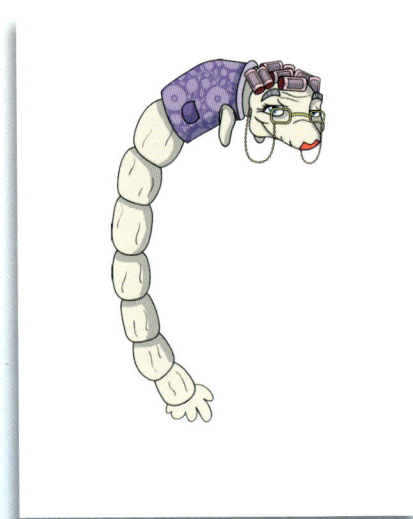

*Proleg (like a stump) near the head
Wormy, segmented body
Prolegs visible at posterior end
Red or white or tan color

*Unlike the caddisfly

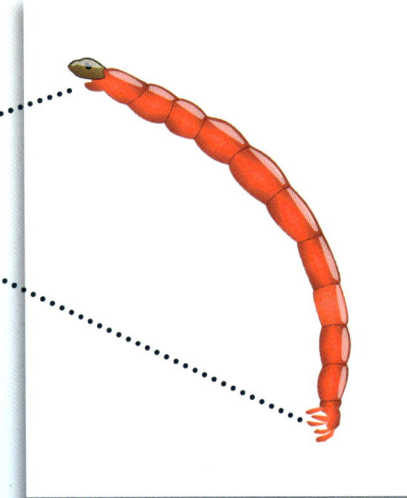

Glossary

Dorsal: Along the top or back of the body, as if from head down the spine (if bugs had spines)

Dorso-ventrally flattened: Appears squished flat, is flattened from backside to front

Gills: A body part allowing the critter to breathe air in the water.

Lateral: Along the sides of something

Proleg: an unjointed mini-leg

Posterior: Rear end (opposite of head end)

Segmented: Made up of visible segments or sections

Telescoping: Extends out like a telescope, getting much larger as it opens

Ventral: Along the inner side of the body, as if from nose to belly (if bugs had noses)

Wing pads: The pre-wings seen on insect larvae in the water. When an insect hatches, the wing pads break open to reveal the new wings within.

The Secret Life of Streams
Extensions/further resources to explore our connections to Loralie's world and learn more:

- *Project WILD Aquatic K-12 Curriculum and Activity Guide,* by Council For Environmental Education: projectwild.org

- *Life in our Watershed, Investigating Streams and Water Quality,* by Sacramento Splash: sacsplash.org

- *A Kid's Guide to Flyfishing: It's More than Catching Fish,* by Tyler Befus.

- *The Streamkeeper's Field Guide: Watershed Inventory and Stream Monitoring Methods,* by the Adopt-A-Stream Foundation: streamkeeper.org or call 425-316-8592 to order a copy.

- *My Healthy Stream,* by Trout Unlimited (TU) and Isaac Walton League.

- Watershed protection, simple steps to protect Loralei and all the creeks, every day! mcstoppp.org/watertest.htm

- Isaak Walton League, and Save Our Streams program: iwla.org

- *Measuring the Health of California Streams and Rivers, A Methods Manual,* by the Sustainable Land Stewardship Institute: slsii.org

About the Author

Lynell Marie Garfield has made her debut into the children's book realm with The Secret Life of Streams! A masters-degreed environmental scientist, Lynell has worked on water in nearly all its forms. Learning first-hand from stream macroinvertebrates is a powerful experience, and Lynell studied creeks with students in the Snake River, San Francisco Bay and Russian River watersheds, translating the world of aquatic insects into a rich understanding of stream health. Inspired to share a fun environmental ethic with children, Lynell has blended creative story-telling with real world fundamentals.

Lynell lives with her husband, Tom, son, August, and Frank the cat in the East Sierra, where she works as a Hydrologist on the famed Truckee River. Amidst monitoring water quality, managing large sets of watershed data, promoting rain gardens and other storm water protection techniques, Lynell chases mayflies up trails and tells dragonfly tales whenever she can.

About the Illustrator

Daniel Devine is an Illustrator and Graphic designer. He has been working in the field professionally since graduating from the Truckee Meadows Community College in 2007. His passion lies with drawing and cartooning, as he loves bringing characters to life.
Daniel currently resides in Sparks, Nevada with his wife and their 2 year old son.

Acknowledgements

For being the container and the contained, for magically dissolving all in your path, for the beauty in each of your forms and endless inspirations through reflections and primordial messages, I celebrate water.

To my parents, for intuiting that a backpacking camp in the middle of the woods on the mighty Eel River would fill my young spirit, thank you.

To friends and students at the Teton Science School, for sharing stream bugs, fire ecology, and the strange behavior of moose around sleeping bags... and Tim Palmer and Terri Tempest Williams...for poignant story sharing and life adventure. Thank you.

To Colleen Ferguson, Jim Harrington, Monique Born, Joanne Hild, John VanderVeen, BJ Schmidt, Tom and Renee Wade, and my extended Yuba family - thank you for believing in me and chasing those bugs.

To Sudeep Chandra, I thank you for our seminars, rich in riddles of sharing conservation ethic with others, and then for booting me into a Kickstarter campaign.

To my wonderful focus group, Rachel Mazur, Jill Romine, Lori Bellis, Mary Cronan - thank you for sharing excellent ideas and friendships with me.

For Willow Tequillo, the New Breed Healers, and Lobsang Samten, thank you for dreaming with me.

For Danny Devine, thank you for bringing the characters to life! For Jason Barnes, thank you for the field guide images. For Manny Becerra and Kelly Peyton, for the journey. For Sandy Silva, for sparkles, margins, and years of encouragement.

For Glenn Miller, for being my very first funder. To all my incredible KickStarter backers, I couldn't have done it without You!

For Charles Goldman, I am happy to have your kindred spirit on this planet.

For Cindie at Lucky Bat Books, I must be one Lucky Bat!

To all my friends, who listened to me and supported my ideas for all these years... you know who you are. I am truly Blessed.

To TQ, for being my life partner, and last funder.

For all Beings, past, present, and future, working with children: My wish for you is that you may always give and receive inspiration, in healthy and unmeasured doses.

Praise for The Secret Life of Streams

Every child is intrigued by water. The question is how to transform that wonder into a desire to learn about, and protect, our aquatic world. In this refreshingly lively and informative story, Loralei the Mayfly takes us beneath the surface and introduces us to that which lies beneath. The reader emerges wet and wild and ready to save the planet.
- *Rachel Mazur, PhD, Wildlife Biologist, Author, Mother*

Loralei the mayfly will capture the hearts of readers young and old. The Secret Life of Streams eloquently depicts life in a stream and elements of the web of life, leaving the reader clamoring to learn more about the special creatures and their world. Throughout this fun and engaging, larger-than-life story, author Lynell Garfield embeds accurate science content... a balance that makes this a must-have for anyone who believes in the importance of expanding the curiosity and knowledge of our children through the exploration of the natural world.
- *Laurie Gray, Washoe County School District Administrator (retired)*

A timely masterpiece that will inspire the next generation to think about the natural world around them. The tales and illustrations will be sure to captivate children into unlocking the beautiful world in which Lorelei lives and the challenges facing our streams.
- *Dr. Sudeep Chandra, Associate Professor, Limnology & Conservation Ecology*
Associate Director, Castle Lake Environmental Research & Education Program, University of Nevada, Reno

Lynell Garfield and skillful illustrator Danny Devine have captured by word and art the exciting life that exists unseen beneath the flowing waters of the world's streams. Her pen and the cartoons provide life, education, and adventure to readers, both young and old as these denizens of the aquatic world move through their life cycles.
- *Charles Remington Goldman, Distinguished Professor of Limnology Emeritus*